Scarborough Fair VIII
AN ANTHOLOGY OF THE ARTS

edited by Timothy J. Allan

*Scarborough Fair* is an annual publication made possible by the generous financial support of the Scarborough College Students' Council, and the Scarborough College Cultural Affairs Sub-Committee.

The editor would like to extend his heartfelt gratitude to the material selection committee for their consummate time and patience. They are Stephen David Michael Kischuck, Catherine Long, Kelly Mansell, Minnie Sakhuja, and Regina Stemberger.

The typesetting and layout were done by Shona Nicholson

Printed by the University of Toronto Press in an edition of 3,500 copies.

Scarborough College, University of Toronto.
Eighth edition, 1980-81.

ISSN 0318-1499

## Maudlin meanderings from on high

It seems entirely suitable to the mind of this humble editor that this little book in your hand at the moment ought to appear in early April. Is not the spring our traditional harbinger of renewed life and creative vitality? *Scarborough Fair*, over the past seven years, has always attempted to complement the mood of the season. Hitherto shrouded or unknown talents among students, staff, and faculty mysteriously appear before deadline, proffering new poetry, prose, photographs, and illustrations.

This year's edition has seen a veritable garden of verses flourish -- and hopefully, not simply a child's one. The response by the community of Scarborough College was nothing short of breathtaking. The total number of submissions to the selection committee is recorded on the last page, and don't expect any favours from me. An excursion across every page in the book is the best way to reach the end.

The end of this introduction is within sight, so the following will be brief. Unfortunately, brevity itself is a necessary part of this book. A fairly good volume could have been made out of rejected submissions alone. For those who found that their work was included in this category, do not despair; we are hardly omniscient, but you are hardly perfect. Together, we may just learn something. Steel your nerve, and use this experience as a steppingstone to others. Remember to autograph a copy of your bestselling book for me when I creep to your door, wretched and penniless, ten years from now.

Finally, thank you to everyone who offered us assistance, advice, and moral support. This book is unofficially dedicated to the memory of John Lennon. Rest in peace, John. We, too are simply trying to give peace a chance.

Tim Allan, editor

## English Cookies

*Lucille Chenier*

Poems are decoratively arranged
In designated anthologies
Like cookies on a silver tray
Various sizes and textures entice
Symbolically decorated
Some sweet, others
Take longer to chew.
A few taste of cinnamon,
Branch of the laurel family
Love of poets.
Pick and choose, but
They are all worth a nibble.
Take some away
Sharing tastes of wisdom.

## Kiss Me, I'm a Poet

*Michael Anastacio*

I meant to pin her for the count
have her succumb to my sensitivity
but she had a black belt in relationships.

## Evening with Cohen

*Rachael Boles*

tonight
a sharp shower of metaphors
caught me as I ran to my car
soaking me to the skin
then I stepped in a puddle
of juxtaposition
and blinded by allusion
missed seeing the tenor
tripped over the vehicle
and hit my head against
a concrete universal

you can keep your metaphors
I am not sure I want to come
this way again

## Ink on Newspaper

*Kelvin Sealey*

Black on white
Small on large
Tangible on intangible
Words on abstract white
Meaning where nothing used
to exist. Murder and death
and people all words on
what was once a tree

## Prologue

*Michael Kischuck*

I know of an existence which has passed my way,
And given meaning to a random collection of events
Which I have often referred to as my life,
But was never happy with.
And I can say my life was never spent,
My happiness is still not as 'twas meant.

## Sonnet

When I think of her, I think of roses,
For a rose is a thorn bed topped with pretty petals,
I cringe at the thought of holding her closely;
She is beautiful, yes, but sometimes it hurts,
And I have to turn away.
I am then drawn back by beauty,
Inevitable to be hurt again,
But for some reason I persist.
Alas, there's no solution to my problem,
Roses to admire and smell,
Are not always to touch and fondle,
And this is my dilemma.
To see, to touch, to smell is what I need,
I reach out to her, driven by my greed.

## Epilogue

What am I to her, or her to me,
That she should put me through this.
How like a beautiful tasting poison she is;
I drink it, but only damned by passion do I thus,
For she is my life and death juice,
And water is scarce these days.
And on I go,
Inevitably to be hurt...again.

## Love and the Sun

*Lee Wallen*

Love is, and should be, not unlike
The rising sun, for it is beautiful and new.
It brings life and colour, and cleanses the air
With its purity and promise.
Its presence is felt often, each time more alive,
More powerful, more breathtaking than ever before.
It affects all, helping us change and know.

So do not take glory in adoration,
Nor feel humility in worship,
For you are a part of the cycle.
And when the sun sets, reflect, do not mourn.
The sun will surely rise upon you again.

*Park Quietness, by Thomas Copeland*

*Photograph by Thomas Copeland*

# Intersection

*Earla Wilputte*

Four thirty in the morning. A lonesome time because it's hardly night and it isn't dawn. It's just black, and even the stars seem to disappear when there's nobody there to wonder at them. Nobody in their right mind is up, though, at four thirty in the morning -- unless they haven't been to bed yet.

Al was up. Had showered, dressed and eaten, in fact. He rubbed the stubble on his face an smiled up at the sky. This was his favorite time of the day. His time. Private and cut off from the rest of the world. He crushed out his cigarette with his boot and climbed into his rig. He smiled again as he looked down from his new height, then he started her up.

The sound of the engine always excited Al. It was more than noise in his ears, it was a rush in his blood and a jump in his gut. When he heard that first swell and the almost deafening growl as the rig kicked into life, he felt like hugging the thing. He's never told anybody that, but it was such an experience to feel that monstrous machine change from a dignified beauty to a snarling animal almost clamping at the bit, he'd swear it was alive. He loved that rig and he loved being alone in her.

He'd tried once to show his wife what he got out of it, driving it and loving it. They sat together, barrelling down the highway -- he'd even pulled the horn for her. He felt like a kid sharing his favorite toy. But she didn't like it, couldn't love it. Al knew she didn't understand it when she'd tried to tell him how much she enjoyed the "ride" in his "truck." If she's just laughed, or clapped her hands, ever stuck her hand out the window, he'd have know she liked it. But saying it? Al knew you just couldn't **say**. You felt, or you didn't. She didn't and he did.

So now he was alone with her. He was alone out of her, too, and that was a different alone--like lonely...But he didn't want to think of that now. Just get her on the road. He figured he'd make Edmonton by one.

He didn't bother switching on the C.B. Didn't want to be bugged by those truckers who wanted to make conversation at four thirty in the morning. Too green. Real truckers didn't talk like that, or at least Al liked to think so. He's been driving for sixteen years, and he'd be goddamned if he wasn't a real trucker.

A maverick. That's what his wife had called him. A cowboy. She meant to hurt him with the names, but she didn't. He liked the idea. Suited him. Hero with no name riding into some town. "Who was that man?" they'd all ask when he was gone.

Al laughed to himself. Dreaming on the job, eh? Fantasizing. All part of the life, though. He liked to drive, his arms around the wheel, him up there and thinking he was something special. Dammit. He was. That little girl in the office sure envied him. Of course, her sense of it was kind of romanticized, but the way she talked about it...He'd take her out in the rig sometime. She'd understand. Shit she was excited just looking at it, and her eyes'd light all up. And when she looked at him, he thought for sure she'd say something, but she only raked her fingers through her hair and shook her head. Yah, she understood like his wife never could, and she never used words like "nice" and "cute." "It must be some kind of life, Al." That was all. But that was everything.

He didn't tell her about its problems, though. The hemmorhoids, the cramps--they'd sure tarnish her dreams. He remembered being carried, crying from his truck, his legs all cramped up, and the pretty nurse shooting his butt full of muscle-relaxant. And the nights he'd try and catch a few hours sleep in the truck. Cold and stiff. She sure didn't offer any comfort when you didn't turn her engine on. Al chuckled. Like any woman.

He drove without thinking for awhile and just watched the landscape. He loved the fall. With the sun coming up, the trees looked like they were on fire or like some artist had gone mad and used the world as a palette for his insane colours. Al shook his head and muttered, "Goddamn poet, you are." He remembered reading once where fall leaves smelled like cinnamon and he wanted to smell a whole face full.

He noticed the bodies of animals that had been hit, lying at the side of the highway, and tried to think of something else. A rabbit, a raccoon and a skunk that stunk up the cabin with its useless defence. Al felt guilty driving his huge, powerful machine, as though he was somehow responsible for each killing. Maybe they committed suicide, he thought. Yah, a bizarre pact with nature, and the animals' bodies were warning signs. Shit better turn on the C.B.

He listened to the garble for an hour or so; where the smokeys were, what pretty beaver would do you for free, a couple of good "gag and pukes" nearby, who was hauling what...Al just drove and listened, not saying anything. All-knowing. Al knowing. He chuckled.

After a while he stopped for coffee and some lunch, watching the frizzy-haired waitress watching all the truckers. She eyed them all and laughed at all their jokes and bent way over the counter when she was wiping it off. Al paid her no mind and so she didn't bother Al with her teasing and her giggles. He retreated back to the rig's cabin to eat by himself.

He was nearly there. He was just approaching the big hill before the bridge that led into Edmonton. Al read the sign "No stopping on the hill" and prepared to move his baby up there. He enjoyed this part; he always felt like he was trying to coax a brontosaurus up that hill. It wasn't dangerous to him now like it had been when he was just a kid -- it was just 'different.' Trying to move a machine where it really wasn't meant to go was a kind of challenge -- one he hadn't failed yet and didn't fail this time. King of the mountain again. On the top of the world.

Starting down the hill he noticed a station wagon, parked, at the bottom. Sitting so serenely, like it owned the whole goddamn hill. Al knew he couldn't stop. With sixteen bloody gears it'd take him at least half a mile to slow her down, even. Just as he figured he'd better go round the car, he saw it. He thought he was dead. Another rig was beginning its assault on the hill. There was no time to think, to move. There was no time.

Al wanted to smash through the damn wagon, but was afraid there'd be a whole bloody family in it. He didn't want to hit the other truck because that would be suicide and Al'd be goddamned if he'd kill himself because some bloody fool had parked at the bottom of the hill. He didn't know what to do. He felt like he was watching a slow-motion moive that wouldn't end. He pictured himself as a pile of guts at the side of the highway, next to a rabbit.

Now he was close enough to the other rig to see

the driver's face, and Al stared at himself sitting in the other cabin. The three-day beard, the plaid shirt, and tightly drawn mouth were Al's. He wanted to reach out and break the mirror. He wanted to cry out and end the nightmare but the eyes controlled him and told him not to be afraid. Suspended in time, Al was calm. Peaceful. Accepting. He closed his eyes for a second and the world began revolving again.

The other trucker motioned for Al to pull into his lane while he chanced the shoulder. It'd be close. Their lives passed.

When Al finally got his rig stopped on the flat, he just sat. Drenched in sweat and shaking, he cursed. He cursed the car, the people, the hill and the rig, then he jumped down onto the road and began walking back to the wagon.

A woman was crying and pointing and screaming. A man lay at her feet, spread out on the road like sweet Christ on the cross. Al tried to digest the scene. Did the rig catch him? Did he pass out? What if he had a bad heart? He grabbed the woman and he shook her. Trying to make her shut up. Trying to shake some sense into her.

"We were only parked for a minute. We wanted to get a picture of the trees. And then the trucks went by so fast and one stopped and the trucker came here and hit my husband. Thank God you came back to help him!"

Al looked up the hill, but the other rig was gone. He looked down at the spread-eagled husband out cold on the road. It was like the guy had read his mind. Al began to chuckle. The lady stared at him, and he laughed even harder. He looked back up the hill for a moment, then he turned to the woman.

"He beat me to it. Shit lady, I came back to hit

your husband.'' he laughed, and walked back to his rig.

He started her up and began driving. Figured he could still make good time getting in to Edmonton. H shook his head. Then he smiled as he realized he'd been beaten at his own mind-game. ''Who was that man?'' he asked himself in a mock tone of awe. And he grinned again because Al knew. Even the little girl in the office would never understand that.

## *Jim Nolan*

He is the thief that comes in the night.
Orcus, Lord of the Shadows.
Thanatos, friend of Hades.
The reaper that harvest a crop of grief and despair.
He is feared.
But to those who know only misery and pain,
To those whose hours are filled with torment,
sadness,
And hunger.
He is often known as friend.

# Helpless

*Dilip Banerjee*

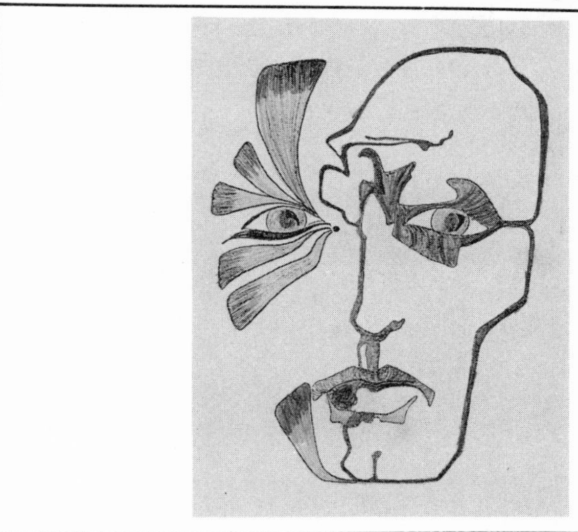

*Untitled, by Evelyn Shah*

Helpless
I breathe, I see,
I eat, I do...
do what...what have
I done?
So many things
wrong
with the world;
pain, hate, hunger,
suffering;
seeing I suffer
and think, and talk
but do...what?
Helpless
I see, talk less
and do
nothing

## The Internecine March

*Sheldon Zelsman*

So I continue fighting
  with my hands up in the air.
The darkened plains are battered,
  and I am unprepared.
My mind is thundering forward
  while I march through the rain.
And the graveyard is open--
  I am shrouded once again.
For I am somewhat older;
  undying decades spin.
As long as I am breathing,
  I know I cannot win.
I'm on an internecine march;
  my friends are all untrue.
I bring the end behind me,
  for my days are very few.

My heart is all but trampled
  with the coming of the days.
I know not where I'm going--
  my name has flown away.
I swim in many rivers
  that run within my tears.
I'll drown in chaotic oceans
  that formed from all these years.
I'll feel the pain of falling,
  or poison will I taste.
For if one runs from fortune,
  then one is quite disgraced.
I've given up on fighting;
  my life can't start anew.
So I bring the end behind me,
  for my days are very few.

I've never known happiness,
  and I think I never will.
It's never handed to you,
  and it's taxed by many frills.
I've found but one small answer
  to the question marks of time.
Although it isn't legal,
  to whom is it a crime?
Give me that one small answer,
  each day I have implored.
For I've been handed calamity,
  and my desires are all ignored.
So let me ask for one thing--
  let me glide away unto You.
I bring the end behind me,
  for my days are very few.
I'm tired of marching onward--
  it's too hard for me to do.
I bring the end behind me,
  and my days are very few.

## Retreat

*Lucille Chenier*

I hid in the north-woods
Behind the glory
of the blazing leaves
Licking the wounds of
my civilized hurts
Until
The loon laughed at itself
from the mirror of the lake.

*Michael Kischuck*

We reunited on a sunny day,
Standing but four hands apart,
Physically separated, yet joined in shadows,
Images painted on the lush exotic green grass,
We talked,
Of long nights,
Beside the fire,
Beside the river,
Beside the hill,
Of our first meeting on a dreary, rainy day,
Drenched by the tears of love,
Of our embarking on an epical journey into
The wonderment of newly-found pastimes,
Of a waltz, a picnic, a frolicking butterfly,
Of soft wind, hair gently caressing my face,
Of sparkling red beverages,
Never in excess,
Of Subsequent warmth and contentment in each other's arms,
Of the exchanging of meaningful dialect, soothing speech,
Of comfort, and fulfillment,
Of beautiful minutes, and gorgeous moments,
Of an unfathomable closeness, yet lack of oneness,
Of the happiness of experience, the joys of innocence,
Of the death of a friend, more tears of a different kind,
Of extremely emotional mourning, and attempted comfort,
Of morning, and much regret,
Of one mistake, and much talk,
Of many questions, and no answers,
Of mutual lack of feeling for what once was,

Of severe depression, and reluctant awareness,
Of springtime in the tropics, and winter at the poles,
Of a sharp ray of sunlight, and the sweat on my back,
Of momentary wind, and overcast skies,
Of thunder and lightning, and hard rain,
Drenched by the tears of despair,
Alas, there we stood, unjoined, and forever to be
thus,
Save for the shadows,
That once were.

## Another Still Pond

*Jim Nolan*

A very still pond
Sends back an image
Untainted by the city,
An image more true
Than that of any looking glass.
It lets us see
Ourselves
As we are.
It stares back in judgement.
It glares at us accusingly.
And we're unprotected
From its honesty.

A very still pond
Is unpretentious,
And real.

Could that be why
Whenever we see
A very still pond...
We look for a stone?

## Rainbow

*Noel Napier*

The crystal clear colours of the rainbow,
Sift through the hexagon of glass.
Colours bouncing off the wall,
To collect and be projected on the table top.
Blues next to green next to yellow next to orange
Next to red, with all the colours in between.
Flickering with the movements of
Objects fluttering through the sunlight.
And as the sunlight fades, so does the rainbow.
Wavering, then finally vanishing.
An illusion that when grabbed for,
Slips through the fingers.
As beautiful as a flower in winter,
As precious as a gift from a child,
As free and brief as a summer breeze,
As wonderful as a child's life,
As enjoyable as watching a family at Christmas,
As quiet as an empty room.
Quite real and visible,
Yet unreachable.
It cannot be captured to keep for a rainy day,
For it lives on sunlight itself.
But there is hope that it will come again,
To be watched and appreciated.

# Vision

*Minnie Sakhuja*

She was blue, the colour of a peacock
Streaked with gray and green and gold.
She walked in mists toward a spray
Splashed from a pond, limpid and cool.
She dove, the waters rose.
She rose, the waters clove
To her body in glistening drops.
Spectral, luminescent, iridescent hues,
Softly swirling, cloudy blues
Veiled her about the while she cavorted
In leafy, spangled bower.
Sun dappled, golden shower
Cascaded on the woodland nymph
Sprinkled with shining glings
Of twinkling, jewelled dew
Sliding, flowing anew.

She glinted in rosy rays,
Sparkled in sunny haze,
And hung suspended in a glittering shimmer;
She shone for a moment in a blinding glimmer.
Then on a final flash
In a trembling tear on a lash
She vanished in the space of a blink
In a tear that fell with a wink;
A vision held in the stare
Of a brimful eye, and rare.

# Persephone Revisited

*Tim Allan*

Raise a glass to the mild mannered experimentalist
Without him I'd never be what I am today
A specimen under a slide
A polar bear sucking the oily ooze
A rat doomed to a treadmill
My open organs exposed to his electric touch
My brain pulverized by his steely rod
My unborn children ripped out
Destined only to play formaldehyde and go seek

Make a toast, will you, to the experimentalist
And here's to you, Prometheus
And here's to you, Sisyphus
Gather round for another round, now
The time may be too late
But you both live on in me
As the gods live on in him

Just toss me into a Pandora's box in a grassy grave
I will be content
To feel the roots of the trees mingle with my limbs
To feel the earth see what my eyes once saw
Let winter's white reign over his bloodied coat
The seasons have fled from the lab to Elysia

# Reflections

*Frank Sinopoli*

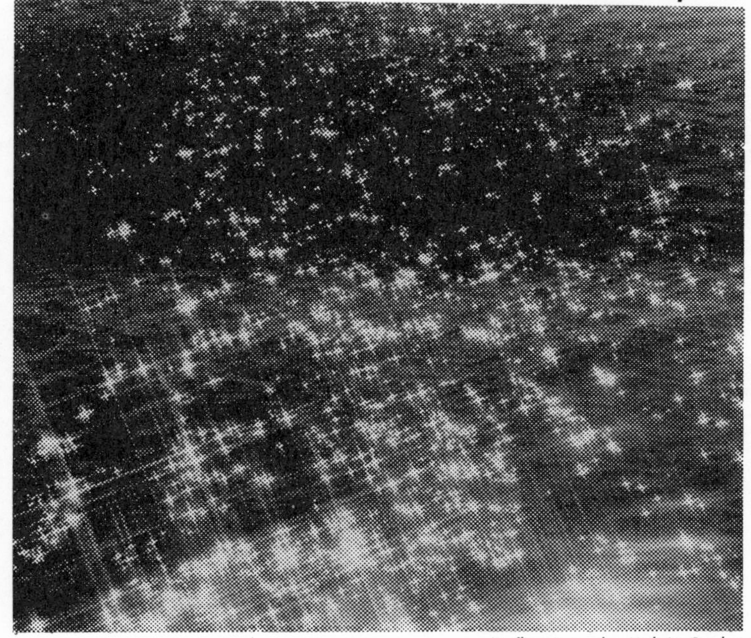

*Reflections, by Kelvin Sealey*

Reflect into the night
The night that becomes light

Memories of childhood
Running through the wood
When joy was the only food

The mirror is still here
Like distant times it stands near

Look into it daringly
With your eyes
And revive the joy
Before it dies

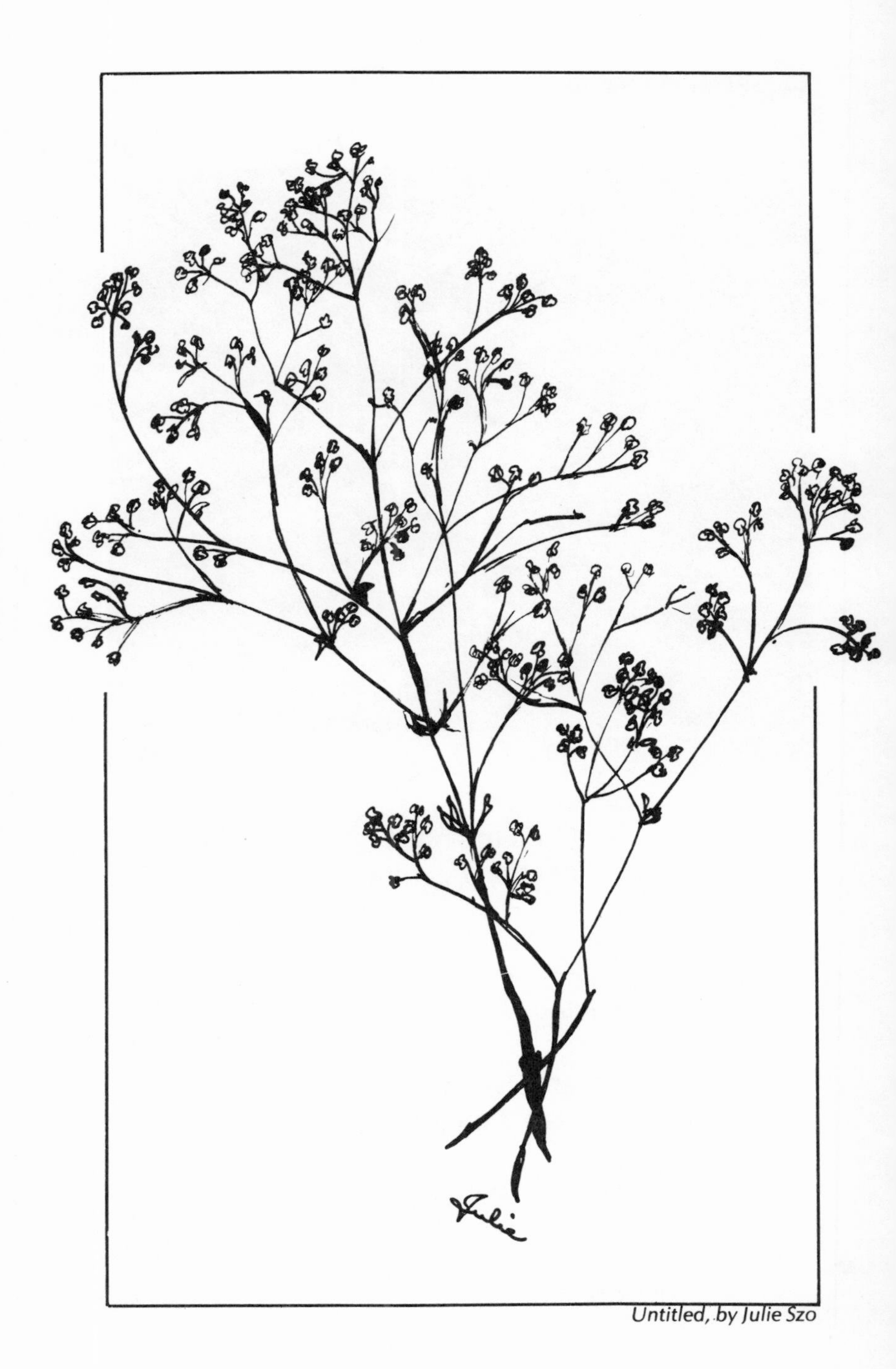

*Untitled, by Julie Szo*

# Student's Progress
# an inspirational allegory

*Brett Zimmerman*

*Acknowledgement must be made to Christopher Marlowe and to William Shakespeare, because I borrowed a few lines; and to some writers of allegories and to Viktor E. Frankl, from who I, consciously or unconsciously, borrowed a few ideas.*

Partway through this life we're bound upon, Student findeth himself submerged in a Verbal Wood, which grows thickly on a Mountain called **Educationus Mons**. Student, and his sole companion Good Sense, are embarked upon an exhausting journey to the mountaintop.

(Night. Student and Good Sense have lit a campfire, and they rest.)

STUDENT: A mighty maze, and all without a plan: the stars shine above us, but I see no stellar patterns, just a random sprinkling of flickering lights.

GOOD SENSE: You must formulate your **own** patterns. Look: there is a lion; over there, a hunter; and to the north, a dipper.

STUDENT: Yes! I see now. And follow the two end-stars of the dipper's bowl and you arrive at that wonderful Pole Star, around which all the empyrean seems to move. Viewed from the mountain's base here, Polaris shines just above the mountainpeak, over the silhouette of that far-distant Tree -- which is our goal. What didst thou say that Tree was called?

GOOD SENSE: "The Tree of Knowledge" -- what else! This is an **allegory**, isn't it?

STUDENT: Yes. Well, let's move on. The sun is burrowing up through the eastern horizon, and it is getting light. Dawn is spreading her rosy fingers --

GOOD SENSE: Cut the crap and put out the fire. Hark! Someone's approaching. (Calls.) Who goes there?

(Enter Pedantry and Dogmatism.)

DOGMATISM: Two servants -- notice I didn't say "**humble** servants" -- who desire companionship. I am Dogmatism.

STUDENT: And what is **thy** name?

PEDANTRY: Say "appellation," not "name." Men call me Pedantry.

GOOD SENSE: Begone! We need not thy pernicious influences.

PEDANTRY (To Student): What is **thy** appellation? And whither art thou and thy comrade going?

STUDENT: I am Student, and we are going to the Tree on top of this mountain. We are searching for the fabled First Degree which, legend has it, is a bronze chain carrying a bronze medallion. This treasure is said to be wrapped around a branch of the Tree of Knowledge.

GOOD SENSE: It is our aim to secure the First Degree, and then to move on towards even higher goals.

STUDENT: My only aids are Good Sense here, and these weapons: this is my shield, called Fortitude, and my trusty sword, dubbed "Will-to-Meaning."

(They continue climbing.)

PEDANTRY: Allow me to offer another weapon: I mean **words**. As a student, you should make it a practice to impress people with a skillful manipulation of words. **Exempli gratia**: never depend on a simple word when a much grander synonym is available. Never say "comely," but "Pulchritudinous"; refrain from saying "cowardly," but say "pusillanimous"; say not "change," but "vicissitude." Utilise as many synonyms as possible - consecutively - in order to emphasize elaborate upon, the meaning. Commit to memory a barrage of impressive-sounding words, like "eschatological," "Procrustean," and so on. You should also get into the habit of **inventing** words; philosophers and literary critics notorious for this. Let them be thy model. Lastly, insert as many foreign words and phrases into thy sentences as you can -- especially Greek and Latin. Learn the following Latin phrases, for instance: "**ad hominem**," ad infinitum," "**cogito ergo sum**," "ipso facto," et cetera.

DOGMATISM: Thou shalt succeed in impressing thyself, if no one else. That's **my** opinion.

PEDANTRY: Behold( What pulchritudinous wench approacheth?

(Enter Distraction, who caresseth Student)

DISTRACTION: Come live with me and be my love,
And we will all the pleasures prove.

STUDENT: I feel an intrusion; an unwanted member reareth its head.
DISTRACTION: Take thine eyes away from yon mountaintop. My abode is just off this beaten path. Come.
(She enticeth.)
PEDANTRY: Slut! Whore! Prostitute!
STUDENT: Good Sense, what shall I do?
GOOD SENSE: You are free to choose your own attitude towards your circumstances. Shall you stagnate with this wench, or move onward? You only live once; will the life **she** offers be fulfilling to **you**?
STUDENT: Begone wench! Take thy leave!
DISTRACTION: Oh intellectual monster -- impervious to all natural affection -- whose passions lie **dead**!
STUDENT: Not dead, just sleeping. Farewell.
(exit Distraction. Student and his companions climb further up.)
GOOD SENSE: Once we jump over this yawning crevice in the mountain's side, we shall be near the zenith.
(They leap over the fissure and finally clamber to the top.)
DOGMATISM: This mountaintop is quite flat and broad. And there's the Tree of Knowledge, not seven metres away. Our trek's a success! And anyone who disagrees be **damned**!
STUDENT: And there's the fabled First Degree hanging from the bottommost branch!
PEDANTRY: But look! What chained, serpentine saurian entwineth itself around the Tree's base?
GOOD SENSE: That horrible dragon is called

Academic Pressure; Student must overcome it before he can procure the First Degree.

STUDENT: With my sword, Will-to-Meaning, and my shield, Fortitude, I shall defeat the monster. I have not climbed all this way for nothing.

(Student attacketh the beast. Good Sense throweth rocks at the monster, while Pedantry and Dogmatism run and take cover. Beast roareth, shooting fire from its mouth. Student retreats, quite scorched.)

STUDENT: Alas! Mine shield is burnt to an ebony hue. Mine sword is cracked.

PEDANTRY: Thou hast other weapons -- words. Throw at the saurian a barrage of verbal missiles.

(Student attacketh a second time. Dragon roareth.)

STUDENT: Pusillanimous! Eschatology!

(Dragon throweth flame.)

PEDANTRY: Latin! Try Latin!

STUDENT: **Nos morituri te salutamus**!

PEDANTRY: "We who are about to die salute you!"?

STUDENT: That's all I know! It's all **Greek** to me anyway.

(Dragon reareth in fury. Student, his Fortitude a cinder, his Will-to-Meaning broken, backeth off, but falleth into the crevice over which they had leapt earlier.)

STUDENT: Woe is me! My weapons are destroyed. My foe is victorious. My allies cannot reach down into this dark abyss to save me. I have reached the nadir of my existence. Out, out, brief candle! Life's but a walking shadow, a poor player that struts and frets his hour upon the stage -- who goes there?

(Enter Despondency.)

DESPONDENCY: Ye should know me well enough by now. I'm tired of toying around with you; these annual visits are too much for me. Here, take this.

(Offers him dagger.)

STUDENT: A dagger! What is the name, appellation, of that dagger?

DESPONDENCY: "Annihilation." Take it.

STUDENT: Oh Good Sense! Where art thou now?

GOOD SENSE (Peering over the edge of the abyss, calleth): Student! Hast thou surged into the world to come to **this**? Remember, even when all other freedoms are gone, thou hast the freedom to choose thine own attitude towards any and every situation. Shalt thou select a pernicious, nihilistic attitude, or shalt thou overcome all obstacles?

STUDENT: I shall overcome! Despondency, take **this**!

(He stabbeth Despondency.)

DESPONDENCY: He has killed me! I am slain! Oh! (Dies)

(Student climbeth out of the fissure.)

STUDENT: This shall be my final assault. My Fortitude, though burned, can still protect me; my Will-to-Meaning, though broken, can still conquer. Academic Pressure, you've breathed your last flame!

(Student attacketh the Dragon for the third time and slayeth it.)

STUDENT: **Veni**, **vidi**, **vici**! Academic Pressure **can** be beaten! And here, the fabled First Degree is mine.

(He taketh the treasure from the Tree of Knowledge and hangeth it around his neck.)
DOGMATISM: Congratulations Student. Pray, why is the medallion **bronze**?
GOOD SENSE: The Second Degree is silver, the Third is golden. Let us move on, Student, for as you see, to the north other Mountains with other Trees and Dragons to be encountered.
STUDENT: Yes. Pedantry and Dogmatism, you may accompany us further; though as I grow older, wiser, and more mature, I must eventually abandon thee. But for now, we shall press onward together.
GOOD SENSE: Let us not lose sight of our aspirations --
The pole star of our hopes and inspirations --
The guiding light towards which we always move,
But will never reach; though we do not want to.
We shall not stagnate, but shall surge on ahead
Like a river to the sea towards our goals.

## For Stephen Forbes Ross

*Michael Anastacio*

Death
hits you in the face
like a misplaced rake.

## Whale Song

*Noel Napier*

Quiet silent sea.
Moving, yet you can't tell,
Silent; a clicking.
Quiet; a cry.
Echoing through the deep.
Fading; another - deeper.
Echoes, and fades away.
Another sound,
In the eerie blackness.
A song,
Going up and down,
In a melody,
Unknown to man.
Singing over miles of emptiness,
Filling the ocean with its sad lonely song,
Trying to reach ears that are not there.
A melody, a song, sad.
Getting gradually louder, more clear.
Soothing, mysterious, comforting.
Then, slipping through the inky blackness, Smooth,
calm, singing, swimming,
Graceful, lonely.
A whale and its song.

*Humpback, by Noel Napier*

## Eden

*Debra VanVeen*

Desert Rose lived in a dry place,
Where the rain never fell
But the sun shone bright,
Lighting all around.

In this place, was no place to hide,
For the ground was cracked and hard.
Trees refused to grow.
The animals of this puckered land
Tucked their bodies and shaped their nails,
To survive this barren place.

Desert Rose stood,
Only three inches tall,
Pink-red blushing hue.
She crouched, and low,
Where snake eats rodent,
She protects the beetle in her shade.

But as the day's heat wore on,
Hour after hour,
She wished for relief from the sun.
For her seed was gypsy, carried far by a storm,
And weary she grew with the land.

Her roots sank foreign through the dust on the land;
She bent and sighed.
And could not she the struggle understand;
She plumbed and tried!

In the garden from whence she blew,
No need to struggle to survive.

As a seedling young
Clear water was supplied by a long and plastic tube,
And her mother, in her beauty, was admired.

The sun in that place was comfort,
A warm potion, that made her mother spread her
petals
And smile up into the sky.

But Desert Rose, now fading from the sun,
Tilted her flower forward;
Could not find tears to cry.

She remained a comfort to the beetle,
A landmark sheltering a probable meal.
But the rain never fell,
And her sorrow and her purpose, were never
understood
By the occupants of this strange land.

## Classical Music

### *Maureen Wilson*

Close your eyes and absorb---
hues of purples, violets, greens
and blues---
        flowing, feeling sensuously
together,
      Arousing a kaleidoscope of sound...
                              emotion
The old masters wanted it that way;
Movements, andantes, crescendos, fortissimos...
                              Classical.

## Oh Fated Man

*Debbie Stull*

Oh brilliant sun;
your day has been well spent
your minutes are few remaining
disappearing behind majestic tree tops
casting your final rays
you wait tomorrow's arrival

i wait with you

Oh sable night;
your silence soothes the soul
your influence is universal
commanding time to a standstill
spreading your blanket of silence
you wait tomorrow's arrival

i wait with you

Oh fated man;
your mind has ceased to reason
your thoughts have inward turned
striving
creating your eventual destruction
you wait tomorrow's arrival

i dread for you

## Another Love

*Brenda Bush*

Until the hurt within has passed
And inner battles have been won,
Until sorrow is no longer cast
On those moments of reflections...
Can there be another love.

When caresses can again evoke
A passion that's been soundly sleeping
The peace of mind of which you spoke
Will softly soothe what once was weeping...
And allow another love.

No success will be as sweet
As moments spent with one who cares
But until you look you will not meet
A happiness that awaits you there...
In the arms of another love.

## the woman

*Catherine O'Toole*

There is a part
that is the child
    gleeful, giddy, so happy,
    so unhappy
all without cause;
        that being cause enough.

There is a part
that is the boy
    tree-climbing, mud-swimming, so proud,
    so cocky
  ready to fight;
        to wrestle with the wind.

There is a part
that is the lady -
    a maiden, a bitch, so cool,
    so surprised
to be the lady;
        to be believed.

There is a part
that is the man -
    strong, broad-shouldered, so free,
    yet unsure
of her manhood;
        and of its place.

These are the segments
that make her whole -
    complete, complex, so much
    to know but

love them all;
    love the woman.

You are part woman too.

## Sleeping Girl

*Michael Goodfellow*

Hello sleeping girl
Features so fine
Don't awaken sleeping girl
Sleeping girl in my mind

How peaceful you are
At ease with mankind
Don't awaken little star
Please remain in my mind

Sleeping girl so serene
You are in my heart
Prettiest girl I've ever seen
We shall never part

## Union

*Tim Allan*

Touching you
I lose touch with everything else
Nothing else matters
Nothing else fits quite so well
Like Aphrodite, a deity from another day
Like Hermaphrodite, you complete me
Made for me alone, my love
My Aphrodite, my Hermaphrodite

## About Hating Sad Goodbyes

*Lee Wallen*

She said with a smile,
"It's all the same to me."
But in her mind, she knew
She'd never be the same.

She said with a frown,
"It's really too bad."
But in her heart, she felt
She'd never smile again.

And as I walked away,
She said "See you later."
But as I glanced over my shoulder,
I saw the flickering candle of hope
Doused by the smallest tear.

## Old Man

*Jennifer Davis*

Neatly tucked away,
Conveniently far from home,
Under the loving care of strangers
An old man sat
Listening to voices he could not hear
Touching faces that were not there.
Look into his eyes, so hauntingly dead,
Brimming full of tears not shed.

# The Loathly Lady

*Suniti Namjoshi*

> *"My lige lady, generally," quod he,*
> *"Women desiren to have sovereynettee"*
> **The Wife of Bath's Tale**

But suppose that Queen Guinevere's Court had said to Arthur, "If it please Your Majesty, 'What women most want' is a woman's question, and it would be more fitting to send off a woman to find the right answer." And suppose Arthur had agreed, then what would have happened? Imagine the scene. Queen Guinevere's on the throne. She looks at her ladies and askes for volunteers. A few step forward, but their husbands object, their fathers object, their children are too young, they are too young, and besides it's most improper. The Queen gives up. Arthur is sorry, but he had expected as much. He summons his knights and they throng about him. He has a hard time deciding which one to choose. He picks one at random. And after a year the knight comes back with the loathly damsel and a suitable answer. The answer's a good one and the men laugh. Then they settle down to a good dinner. Nothing is changed, no one was hurt, and even the knight's satisfied because the loathly damsel is changed overnight to a beautiful woman. Chivalry flowers. They are all of them gallant, and have shown some concern for the Woman Question.

## Complaint

*Suniti Namjoshi*

Two knights in a forest. It's early in May. Bright sunlight filters through the leaves. A damsel in distress is weeping quietly. One of the knights has abducted this damsel. The other is her lover. The knights are fighting. Her lover wins. But the problem is that the damsel in distress has already been raped. The knight, her lover, is greatly distressed. How can he marry her? He grieves bitterly.

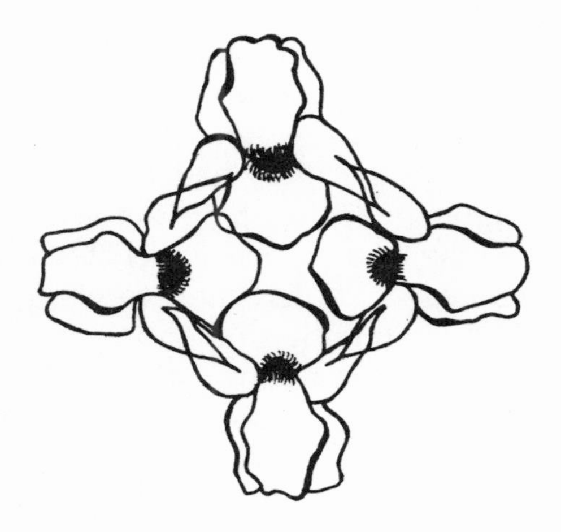

*Untitled, by Amy Morton*

## Broadcast Live

*Suniti Namjoshi*

The Incredible Woman raged through the skies, lassoed a planet, set it in orbit, rescued a starship, flattened a mountain, straightened a building, smiled at a child, caught a few thieves, all in one morning, and then, took a little time off to visit her psychiatrist, since she is at heart a really womanly woman and all she wants is a normal life.

## Snow

### *Patrick Wong Shue*

On a calm winter's night
Not a whisper can be heard
Fluffy flakes float graciously to the earth
captivating the heart
beckoned by the land
So soft, so smooth, so peaceful
a virgin in her purity, her elegance
so white, so pure---unified
a soft light caresses the tender moment

The moon echoes day
unwillingly, day awakes
virgin snow no more
purity and elegance lost to human existence
roughened and blackened by man's grease and
grime
virgin innocence, virgin beauty
slaughtered unknowingly, callously
gone...gone until the next snow
until the next moon.

*Old on New, by Kelvin Sealey*

## City Strip

*Tim Allan*

Pale pearls gleaming round an icy swath of flesh
Madly piercing the smoky darkness
Winking slyly to entice
"Bend me, shake me, anyway you want me,"
And the offers come
With smirks and knowing chuckles
And furtive glances left and right
Relax, he said to me
As I spilt my drink on my lap
Life in the fast lane just ain't possible
Unless you use the passing lane once in awhile

"Come back to fill up the tank again?"
She laughed a little too loudly,
"Why don't you change gears
While I shift into something more comfortable."

Life on the road or in the streets
Or on the outskirts or beneath her skirts
All seems the same under these cold city lights

## Canada --The land of golden opportunity?

*Jacqueline McKay*

"Your time's up Mac."
After this, the only words exchanged between the two, Sally rolled over to the side and sat, on the corner of the musty smelling bed and started to pull on her nylons systematically.

"Hey, I **said** your times up, now come on, I got

other business to do!"
With that, she put her money into her garter and strutted out of the room.

"OK Joe, room's free now, here's the key."
The fat, stodgy manager, never looking up from his paper, merely held his greasy palm out enabling Sally to drop the key in it on her way past. As she stepped out onto the street, she was momentarily blinded by the gleaming sunlight. She blinked twice and recovered from the rapid switch of dark to light.

When she had reached her perch, she stood in her familiar way, one leg out to show the merchanise with the hips thrust forward to entice, and started to think.

She thought of many things while the time flew by. Mostly, she thought of what she would do after she had finished saving her money. "Perhaps," she thought, "I could get a job in a fine dress shop..."

"Yes, may I help you? Oh yes, that's our new line of Ports clothing just in. The skirt? Oh, it's only $200. you'll take one? Yes it is nice isn't it, I have one just like it at home."

Yes, and her home, it would be nestled in-between the tall oaks of Rosedale, very spacious and open with lots of plants and sunlight pouring through the windows. And clean? Oh, she would keep it so clean, you could eat off the floor, not that you would want to for the mahogany dining table would be waiting laden with food in front of the roaring fireplace made of marble.........................................

"hey...**Hey**, you "open" so to speak, for business?"
Slowly, Sally's head turned around and confronted her next trick, a bleary-eyed face with the blood-red tongue darting in and out of dry, chapped lips.

"Ya, Mac, give me the forty bucks."

Slowly, they made their way back to the dingy hotel room with the musty smelling bed and the neon sign "ROXY" blinking on and off, lighting the room with flashes of red and yellow in a constant rhythm.

## I Saw His Bones In a Photograph

*Michael Anastacio*

I saw his bones in a photograph
piled up like unwanted trash
brown and fossilized, staring sadly
as if his eyes were hidden there
as if they were saying: "Remember me?
I held you once, I bounced you on my knee
chased you in the garden, back in the old house
back in the old country, you followed me in the
parades
remember? I played in the band,
you were only four but you marched
head erect and proud stamping behind me
and Sousa never felt better.
And that horse, do you remember?
That wooden rocking-horse I made for you
I can still see the expression on your face
when I brought it home, you nearly tore my face off
with your hugs. And the time we crossed the islands,
surely you remember, sitting on my shoulders,
crossing the bridges and cliffs, you cried
thinking we'd plunge to the depths below,
I laughed as you cried and your heart was jumping
and you were only four years old and I loved you.
Do you remember? Surely you must. And the time
I took you to Jose's tavern and gave you wine
from a tall glass, and you, the mimic

calling for another round and you were only four years
old
and God how I loved you. And soccer games where
you pretended
to know the players and the rules and the score
but I knew you didn't care and just wanted to be there
with me, an old man with thick hands. Remember?
I know you do, those nights candle-lit and cold
when I'd tell you stories of witches and sea-serpents
of princes and kings, and how you snuggled close, eyes
in wonder
and you were only four years old and I loved you
and at the airport, that final night, my eyes were blurry
it must have been the wind, and you were on that plane
and I waved. Do you remember? Of course not, how
could you,
you were only four years old and I loved you."

## Morning Ramper

*Dave Ranson*

Seven AM at Coronation Park,
The sun in corner of the sky.

I'm watching the boats rock gently at their moorings;
I'm hearing the seagulls cry.

Before the rest of the shift comes on,
This mount gives release.
Before dealing with the jokers on wheels,
The world and I are at peace.

## Along the Winding Path

*Colin Barnard*

Autumn hail strikes the dead fallen leaves,
The sound of applause surrounds me,
It's everywhere,
And as I bow,
I soak up the acclaim,
And then realize,
That the only thing I soak up,
Is rain.

The sound of an appreciating crowd,
Soon evaporates,
And dies,
Like an echo in the forest,
Leaving a man alone,
Along the winding path.

*Michael Kischuck*

When I think of what it does for me
I cringe,
Stimulation in all ways except
intellect.
What is stimulation to the passive self, or so it thinks.
Why do I feel thousands of emotions pent up
inside,
Unmoving, undeveloped, unknowing self,
Unconscious of what is real
for me, to me.
What am I to it, or it to me
That it should put me through this.
Mechanically inclined, and inclined to act
mechanically,
I can follow the motions
That's easy.
Why
do
I.
I do, why?
Conditioned to accept, I accept the conditions.
NO!
Stimulate, it can. Need, I don't.
Not of that kind.
I've been told. I know that.
She told me.
She has what it doesn't
What is it?
Think about it.
Think about "it".
---------i-----t---------
I can't!
I cringe!

Stimulation in all ways except intellect. What is........

## Canadian Shield

*Lucille Chenier*

Valleys deep
Nippy frosts
Flowers sleep
Summers lost.
Granite outcrop
Abandoned mines
Timbers drop
Sawyers whine.
Northwinds brisk
Wits sharpen
Footsteps crisp
Days darken.
Deer runs
Oiled guns.

## silence

*Ed Camilleri*

single house upon grey fields
lonely voices call
tepid rains beneath the sky
drops of silence fall

dying child in mother's arms
shadows on a wall
fading light from candle's glow
silent raindrops fall

lonely house beneath grey skies
shadows on a wall
endless rains that patter on
drops of silent call

*Untitled by Tony Westbrook*

# Beaches

*William Wen*

This is the story of a beach. It is a story that requires that you, the reader, listen. You must learn to hear the ocean breeze as it sweeps across the sand; learn the timbre of the trees. Do you see the spider sitting on the sand? Do you hear the stillness of the dawn?

About five years ago, a man returned to this beach. with his sandals in hand he moved towards the rushing waves. He stopped only when he felt the sea touch the soles of his feet. He had stood there as if he were waiting for someone. The only witness to all this was a small, somewhat large, spider. Perhaps. like the man, it too had felt a desire--a need--to pause and listen to the quiet of the rising sun. The man was not one of the locals; yet, standing there, by the water's edge, he seemed to belong here. Perhaps, it was his relaxed manner; perhaps, it was his smile. Th man had stood there that day listening to the silence; it was as if he was trying to draw his very life strength from the silent sand about him.

Ever since that first return home some five years ago, the man has been here often. Some say "it's just his hideaway, his favorite vacation spot, you know." Like the spider, a few know better. It is certainly true that people have their favorite place--be it a suite on the Las Vegas strip, a beach house on the coast, or a cabin in the Canadian Rockies. People say how calm and peaceful the man seems whenever he is here; "He's so entranced by it all here when he looks out over -- well not looks out..." stammers one woman. A few think he has gone off the deep end. The man himself, only smiles politely and nods his head--for

by then, he's already lost in his own thoughts and feelings.
It was those unspoken feelings that brought the man back five years ago to stare, with the spider, at the ocean and the dawn.

About twenty-five years ago, it had been a woman with whom the man beheld the beauty of the sand. The sand had been warm then. The trees--full and fragrant--had whispered softly while the leaves had rustled, playfully, in the wind above the young couple. The sand had been warm and inviting as they marched barefoot along the water's edge. The beach had, graciously, accepted their early morning intrusion; the sea flowing caressingly upon the sand.

The sun had shone upon the woman, highlighting her oval face, her delicate and lithe form. Their eyes had shone with youthfulness and dreams then. By the light of the dawn, they found love as the glistening sand beneath them became a reflection of the heavens above. Their dreams and hopes were added to those already scattered upon the sand, dreams of an era.

Yesterday, the man stood upon the sand as the wind beat at his face and arms. His fists clenched. His face tightened. The warmth and the humour, normally seen in his eyes, were gone. The spider heard a low, harsh grating sound, like teeth grinding together.

The man had returned twice to the beach since that first dawn. The first time, he carried a tennis rachet and wore a new shiny wedding ring; the second time, he carried his AK-10 in the fast-fire mode. The ring was in the right pocket of his flak jacket. His movements were no longer those of a young lover as he ran along the shoreline that night.

A spider had scurried across the sand to avoid the marching men.

The man had known that ahead awaited the enemy; but there had been an initial paralysis: his body and mind had refused to unite. The training that had mad the man into a deadly instrument of death had struggled momentarily against images of the sand. His mind created the silhouettes of two lovers walking across the water's edge. He had longed to feel the warmth of the sand's touch, but he had found only the hard leather of his boots beneath him. Even the trees had stopped their whispering; they had only rustled with the sounds of snipers. The man cried out as if to halt this destruction of paradise just as the first grenade exploded. The noise had been deafening, terrifying. Trees and flesh were pulverized. The bang-bang of pistol fire became the rat-a-tat of high-powered assault rifles. The sound of coherent commands became the horrid screams and groans of men as bullets ripped into arms and abdomens. Newly formed craters filled quickly with tree fragments and blackened bodies of friend and foe. The man had never seen death and utter annihilation on such a scale as this. He had used his knife when his ammunition ran out. He had fought the enemy until that burst of metal tore flesh and bone apart. He had fallen upon the sand, the reflection of the heavens--a reflection marred by death and blood. His men, those who were not dead or dying, found him lying upon the sand. He had been blinded.

The enemy is gone. The bodies removed and the craters filled. The tourists and the lovers, like the man, return. Some come to see the site of the bloodiest battle of the war. Others come back to pick up the dreams and the hopes left behind in the sand.

Have you learned to listen to the wind and the sand now? Pause. Feel the warmth of the air. I have told you about this beach now--it's history is probably echoed around the globe. Would you do me a favour, now, and tell me; does the sand still glisten? You see, the doctors won't know until next month whether or not I'll see the ocean again...Until then, do tell--does the sand glisten?

## Cycle

*Won Yip*

My love
when did you stop believing in me
in the feelings that entrance the world
in yesterday's memories
and tomorrow's dreams

My friend
when did you stop listening to me
when I needed someone to talk to
when the world was spinning
much too fast

Stranger
when did you stop noticing me
in the mass of tamed faces
when I desperately tried to reach out
and say hello!

*Photograph by Wayne McCloskey*

# The White Horse

*Wendy Strickland*

oh griff i once could touch your mane
and leap upon your back
to fly through fields of woolly lambs
and laugh my trembling bliss
oh griff we'd win all prizes true
and happy we were free
to gallop past those wretched mounds
of meaning shroud in shades of grey
oh griff we's dance across deep seas
and soar above vast gulfs
no thing could curb our mad delights
no one could quench our glee

but now oh griff I cry to touch
your softly white free mane
I long to feel your supple back
uplift me to the skies
but griff my reach is short of long
in anguish I now fail
I cannot see your shining mane
when lost in mounds of grey
I cry out all your holy names
and count upon my beads
but griff beloved you're drifting far
in clouds of make believe

# Dialogue

*Maureen Wilson*

They spat angry words across the room at each
othere
''I'm going to move out'', she flung.
''Just go right ahead'', the mother fired...
''You'll realize the heaven you had only when
you've face the hell''.
And so the vicious, nailhard darts were directed---
straight to the heart...Their only course was to sting...
hurt.
Flickeringly the high-strung emotions calmed...
died.
For this was only a routine battle of parent vs child.

*Nick Paraschos*

We have seen Love,
We have seen Hate,
We have seen Life,
We have seen Torture
On a desert highway under the sun,
We have seen Pleasure
On a summer night having fun,
We have seen Pain,
We have seen Sickness,
We have seen Fever
On a backyard with our actors' suits on,
We have seen Death
On a lonely mountain from a cliff hanging on,
We have seen all these
And only two will remain:
Life and Death
Passing before our eyes at our last breath.

## Histoire D'Amour en trois actes

*Anonymous*

I

Y avait un oiseau solitaire
Qui aimait la nuit, le vent et la mer,
Un oiseau decu, un oiseau tetu
Mais surtout un oiseau perdu.
Or, cet oiseau blanc, qui lui tendait la main;
Tendresse, espoir et amitie
Sont tous grandis guere sans dessein.
Oiseau blanc, oiseau blanc, ne t'envole pas,
L'oiseau bleu a encore besoin de toi.
L'oiseau perdu c'etait moi.
Merci,
    oiseau blanc,
            pour ta main.

II

Comment jamais te dire?
Quel mot employer pour te decrire?
Toi qui me fais sourire,
Et me donnes tant de plaisir.
Il n'y en a pas.
Mais quand je pense a toi...
  Sortileges d'enfant
  La joie d'apprendre
  Caresses d'amant
  Si douces et tendres
 Un moi-mene retrouve,
 Un toi-meme a explorer.
Si par hasard (mon oiseau blanc)
Tu t'en vas demain,
Ce que tu m'as enseigne
Je vais toujours garder:

Avec toi j'ai appris a aimer
D'un amour sur, droit et devoue.

III

"...L'amour passe comme un orage, puis la vie, de nouveau, se calme comme le ciel et recommence ainsi qu'avant.
Se souvient-on d'un nuage?"
Maupassant

## C'est pour la Vie

*William Murray*

Les ames se rencontrent, s'embrassent
ensuite est-ce qu'elles se separent
Chacune de voir l'autre mourir
n'y pensant guere
Chercher une eglise
est-ce qu'il le faut...ou je l'ose
Je t'aime aussi fort que le soleil
qui ne brille que pour une rose

## Love Is (A Definition):

*Sandra Ward*

Love is the feeling of strength,
The essence of purity,
And the knowledge of perfection.

## Stardust Dreams

*Sheldon Zelsman*

Why can't I grace the stars like everybody else;
why does misfortune cast shadows on my way?
Though I have bemoaned my own destruction,
my blood is thin so my soul must pay.

I look around me and I see the others
climbing up to their destinies everywhere.
The view from the top of the highest mountain
can be yours only if you get there.

But for me I have seen the night coming,
and it has started to come for days.
I cannot live in a world where I cannot be,
where I must practice the blackness of worldly ways.

They say that someday I will stand tall
in fields where my dreams shall have grown.
I believe that I am standing with my own persuasion
in a forgotten field and utterly alone.

In my life I never asked for any of this,
but I am what I am and always will be.
Somehow I will survive through the affliction
surmounts,
but will the stars ever shine on me?

## Friend

*Won Yip*

When I was out in the cold
    you took me in
When I was down and saw no hope
    you led the way

and in the dark when I was scared
    you protected me.
So how do I repay your kindness
I take your money when there's none
to spare
I lie and turn you against once
trusted friends
I eat your food and abuse your children
when you're away
I steal what I want and break
what may be precious to you
I am your friend...

## Exile

*Ed Camilleri*

waiting, dreaming
hopes to share
gentle raindrops
no one there

searching, calling
evening sky
silent snowflakes
drifting by

hidden whispers
lonely ways
reaching, touching
yesterdays

needing, giving
wanting care
fleeting shadows
no one there.

*Untitled, by Won Yip*

## Stagestruck

*Tim Allan*

Tipping back my hat to a jaunty angle
Swimming in the glow
Simmering 'neath the paints
Taking just one more curtain call
As the usher boys flow to the doors
And the house lights come up
And I look up
To see you
Writing
A review
For the morning after papers

Cliched performance she said
And this play's closing after one night

## A New Love

*Maureen Wilson*

I adored him
  --And he knew it
For nigh a year ignored every
other guy I did
  --And nought he thought of it
With love-tinged notes,
passionate glances, I fed his
ego
  --And for granted he took it
Slowly I molted--the notes ceased;
indifferent looks replaced lovesick ones
  --And to be sure, he noticed it
He came on strong; I played coy...
Oh, but I knew when to stop
  --Because you see a new love I had found.

## Defeat

*Sheldon Zelsman*

Who can I blame for my insufferable defeat
when it stems from the inner self?
I've locked all the rooms where I used to study,
and left my mind on a neglected shelf.

How many times have I heard your songs of glory,
and watched as the parades passed me by?
Why can't I learn but a single Hymn of Praise
no matter how hard I try?

Don't tell me I'm not cut out for certain things
when ignorance burns in your eyes.
Your sycophants may take you to the top of the
crowd,
but how lonely you are beneath your disguise.

Don't tell me of friendship everlasting--
my passion is left undone.
So cast me adrift and watch me sail away,
and forget me one by one.

I think that I should let the dove fly away--
captivity crushes her wings.
How can a vagabond offer her a home
when he is always wandering?

I see my life as being completely desultory;
it's hard to wake up to the dawn.
And my ghost sits and laments on the wings of the

dove
who's too haggard to carry me on.

How long will it be before the dove will fly away,
and drop me deep in the hollow?
Me, I can only question the wizardry of Light,
for it's Darkness that I follow.

Don't tell me of friendship everlasting--
my passion is left undone.
So cast me adrift and watch me sail away,
and forget me one by one.

Don't tell me that I'm so greatly introverted--
when you're hurt I'm always around.
But you don't want to know me when I am
harrowed,
for you are nowhere to be found.

Don't venture too far for you'll soon be an enemy;
I'm certain that I'm not mistaken.
The loss of your friendship is no grave matter,
it's just a miscalculation.

The key to unlock my spirit lies within the Truth--
the Psalm that I wrote long ago.
No-one has ever tried to search for the meaning,
and no-one really cares to know.

Don't tell me of friendship everlasting--
my passion is left undone.
So cast me adrift and watch me sail away,
and forget me one by one.

## Bohemian

*Debra VanVeen*

A child fell,
A woman will rise,
Up through the faithless eyes
That circled the ground
On which my wounded spirit
Lay.

Out from a bog,
Running along the open sea,
You will find me;
Unravelling and inexhaustible,
Spanning your face,
For the uncoloured truth.

## L'Amicizia

*Frank Sinopoli*

Mentre il viso scuro,
L'illuminato da due occhi
Chiari, ti guarda a lungo

Le ginocchia si flettono
per far avanzare
I callosi piedi.

Eppoi le braccia ti
Accettano con un
Abbraccio, dolce, dolce.

## Mountain-Climbing

*Catherine O'Toole*

it is difficult to live
within sight of these mountains
and not wish to measure
my height against
theirs.
I see many bases --
and then many peaks.
    (a queer jest of His allows
few to see all), so
semi-blind,
we choose.
which one shall be my forte?
the choice is as enormous as the consequence.
some are
    soft;
grassy-greened and leavened.
places of wildflowers, shepherds and rain.
some are
    terse;
granite-greyed and chiseled.
places of rock-water, wise men and wind.
one is
sun-iced and glowing at daybreak:
    another lies misted and lost from view.
one is
orange-kissed and peaceful at twilight:
    another is blunted and changing in hue.
one is
mine
you know,
I have never been this close to the sky before;
but for a lady raised on hills and plains
mountains are just one more thing
to embrace.

## The Dream

*Michael Goodfellow*

Alone,
Running through the wet drizzle and fog,
Along the cobblestones,
Between the dark, dreary, cold walls of the alley,
I stop in the shadows to catch my breath.

Distantly,
Footsteps are following quickly,
Coming closer and closer,
They splash in the wet and are absorbed,
Only to emerge and continue.

Running,
Again, constantly quicker and quicker,
Around the corner I am exposed,
To dim lamplight from a bulb upon the brown brick wall.

Hurriedly,
Continuing through the mist
I come face to face with a door.

Entering,
I am immersed,
Into a room of mumbled voices and clinking glasses.

Escaping,
Into another room, I am warmed by the air,
And all my friends are there.

Awakening,
I can vaguely recall the fear, the hope, the warmth,
Of the Dream.

*The Photographic Machine, by Thomas Copeland*

# The War

*Kelvin Sealey*

The last battalion fell into place behind the ranks of its brothers -- twenty-million strong. It was rumored that this war would be the war to determine the strongest in the universe. But the soldiers who would perform the actual fighting were neither anxious nor apprehensive. To them, it was something that simply happened, and soon was over.

The sun had come up that morning just as it had millions of mornings before. The grass that covered the universe was not especially wet or dry. All things were as they should be -- all things were as they were expected to be.

In the other camp, far away on the other side of the world, things were not terribly different. Their sun had risen at the same time, there butterflies and bees still fluttered through the early morning air, unaware that in a short time, a very short time, a war would be fought--a war so immense that it's magnitude was unfathomable by most of the creatures in the universe. Yet the troops in both camps were not disturbed in the least.

Finally, the Imperial Leader, the Queen of the belligerent forces, climbed to the highest point on camp grounds and spoke to her followers. The clicking and chewing stopped and all eyes were focused upon her magnificence. Her voice resounded through the ranks. She gave commands to all troops simultaneously, for all were equal in this war save the Queen. The advance was given and the army moved, its motions instinctive, its weapons innate, in the direction of the enemy.

The engagement was smooth and inevitable.

Troops engaged troops until the dead outnumbered the living. The destruction of the environment was minimal, limited to a few grass seedlings and some wandering non-combatants who happened to be at the wrong place at the wrong time.

In the end, just a few hours from the beginning, the victors claimed their new territory. A messenger was sent to the Queen so she could share in the glory of victory and set up a new home.

The whole bodies of the defeated were few, for most were ripped apart. Their parts would be left to enrich the ground so that taller, stronger grasses could grow to fortify the new boundaries.

Dawn found the victorious troops neither elated nor sorrowful. They had fallen back into their daily routines. The workers gathered the food and fertilized the Queen's eggs; the slaves built and cleaned the passageways so vital to the cities, while the Queen laid her eggs and contemplated when her troop would have to engage the ants to the north!

# Dream

## *Tom Yip*

I am walking through a park with a chilling sensation down my spine. The cool air is crisp and the trees shadow over me. But still I walked toward the dim light. I look behind me. A tall figure stands cloaked in midnight black. Half dead, it approaches me. I increase my pace until I begin to run. Glancing back every second my heart is pounding. My chest tightens and my throat is parched. I steal one last look at it before the hand reach for me...I am falling unsuspended in space...endless...helpless...

## I Walked Upon the Highway

*Michael Goodfellow*

I walked upon the highway
To where the pavement meets the sky
On my way I met a blind man
With a darkness in this eye

He told me things I'd never known
But had never seen before
He promised a world of riches
And held the key to unlock the door

The heat was hot and the sun was strong
It played tricks upon my brain
I closed my eyes and fought the urge
But those blind eyes stared again

For a fraction of an instant
I saw the fireballs deep inside
He was an advocate or the devil
Which one I could not decide

I walked back upon the highway
From where the pavement meets the sky
I said "no thank you," to the blind man
With the burning in his eyes.

*Untitled, by Melissa Cheung*

# The Boy I Left Behind

*Michael Anastacio*

I wore a smile when I was young
and my hunger knew my mother
and her giant apron wrapped my appetite
and I ran and ran and climbed and climbed
and tumbled into the first light of my first sun
and green was my tongue and green was my heart.
I wore a cross when I was young
and the thunder knew my fear
and the black-robed sisters danced in my naive life
they sang as my cross shone and the trees wer tall
and birds simply flew and my faith was new.
I wore a sly grin when I was young
and the blue-jeaned ballerinas knew my meaning
and I was growing and felt strange feelings
and those pig-tails were tempting and I pulled them
and I kissed and blushed and ran

and my heart was red.
I wore a tear when I was young
and my heart knew my love
and both knew the first hurt of the first fall
and my faith grew small
and the white-collared holy men drummed
the first song into my drifting ear
and I was lost, though in their grasp
and my heart was bruised by the first question.
I wore a strait-jacket when I was young
and the wind knew my restless urge
and everything was chrome and beer and flesh
and my legs felt the rope of my parents' time
and I squealed for the first highway
and I laid with my first sin.

I wore defiance when I was young
and the winter knew my lonely hate
and I stormed through the gates of advice
and swallowed every brother's lie
till the chains I used to carry freely
dissolved in the heat of my own quiet rebellion
in conformist truth of a needing compromise
I whistled and I sang and I danced
as I was taught
when I was young and wore a mask.

## Haiku

*Brian Wilson*

When the breeze is right
They rise like an orange cloud,
The monarchs fly south.

## Uniqueness

*Jennifer Davis*

A solitary drop
            crystal clear,
Clasps a single blade
            in a sea of grass.

Encumbered by its heavy load,
The drop is forced to fall
And is swallowed by the thirsty earth.

*Untitled, by Tony Westbrook*

## Widow

*Won Yip*

You
are so special
to me
so when they lay me
down to rest
please don't feel
sad
because what we shared
will always
warm you
even in the cold
lonely nights
when you turn to me and
I'm not there
when our memories
taunt your heart
like obsessed habits
Be still
my love
for my spirit
will protect
you.

# Conventional Hypocrisy

*Michael Kischuck*

"And so, I end my address with one final word. I'll tell you a story. Think about this. A friend of mine was once out doing his job, as some of us are apt to do at times." (General laughter) (Long pause)

He slowly scanned the crowd.

"That's what I loved about him. Peter was a great person. He's dead now." (Complete silence) "When I was awarded the presidency of this company, my first choice of assistants was Peter. It was upon his foundational solidness that my empire existed." With this he proudly adjusted and tightened his tie.

"Yes, Peter was truly a great person." (Reflective pause)

He slowly scanned the ceiling.

"As I was saying, I had sent Peter out to Rio to put the lid on a big deal that I had opened not four months ago. He masterfully pursuaded the Brazilian bastards to give in to my opening offer of five-and-a-half million." (Applause)

"Not having left Rio immediately, a price was put on Peter's head by some resentful employees of the company with which we were dealing. They're all quite loyal to their leaders, and strongly felt that Peter had cheated them somehow. A bad deal with the company would have meant a slight adjustment on their pay cheques." (Sarcastically) "That would have been a disaster." (Slowly and thoughtfully) "They killed my Peter in a back alley in the slums of Rio. I'll never forgive the boys in Rio for that. I'll not deal in Brazil anymore!!" He slammed the podium with his fist. (Applause)

"Peter's gone, but not totally needlessly. He landed the largest deal ever handled in the company, and it's because of his efforts that you will, from this time forward, receive a few extra dollars in your pay cheques each week. Thank you, Peter, and thank you."

## Oh Jailosee

### *Wendy Strickland*

What's the gisting of Jailosee?
I assed the crumblem crowd
Why doors it slam and green at me
and crunk me toode floor?

So Man of Slurbings approached me
And hooked into me cell
He tolt me strike the Booge of Lites
Me anslurbs would afire

I struck the Booge with hooked heart
And hasty hebled breath
But ah a larse
Those burning slurbs
Just fired me cell on smoke

Oh Jailosee doth blind me dune
No freebie chile dam I
De Booze of Slurbs
Has parsed me bye
In brimstoned cells I smolders

## Cobalt

*Lucille Chenier*

Blackened frames
hunch over
empty claims
corporate cover.
Mask faces
consumed by
silver traces,
won't dies.
Springs bubble
thick sludge
heavy rubble,
Gods judge.
Man bungles
Nature struggles.

## Smokestacks

*Kelvin Sealey*

Standing like so many brick giants, they silently release gas we find dispensible, with perpetual regularity. Their expulsions: the odor, the mess, the white, the brown and black, mesh; mingle; combine with one another above us and wait, getting ever stronger and more powerful, till one day it will come back down among us to carry out its predestined and inevitable function. But the smokestacks will survive.

## For A Friend

*Sheldon Zelsman*

How do you thank someone who has opened your eyes
to the past and present treasures of knowledge?

The words of my songs
come out jumbled and wrong,
and I just cannot say what I mean.

All that I've told you, you've read in my words,
and I'm sure you know what I'd like to say.
Though you may not have understood me too well,
it really doesn't matter anyway.

For I am what I am, and always will be;
to err, to succeed through the years.
Though you cannot grant me any of my dreams,
you've helped me more by understanding my fears.

All that you've taught me, I will never forget,
still I won't keep it all to myself.
Because the only way I know how to thank you
is by passing your teachings to someone else.

## A Boat

*Noel Napier*

A tiny boat on an empty ocean
Surrounded by seagulls and jellyfish,
Drifting with the wind and waves.
Abandoned.
Floating lazily,
Rocking,
Back and forth, back and forth.
Quiet rolling waves slapping against bow and stern.
Sails, offwhite and limp in a non-existent breeze.
The sun reflects off sea and polished brass.
Sparkling stars of the daytime.
A cool salty scent kisses the air,
As the boat drifts on,
Forever amid seagulls and daytime stars.

## He speaks

*Dilip Banerjee*

Dear Earth, why do you weep
just because I rest in peaceful sleep.
In all the time I roamed the earth
you gave me little cause for mirth
and when for love I sought
I found for a fee could it be bought.
In vain I sought to see some light
and in the end discovered fright,
and all your wars, and cruel hate
did not my growing fears abate.

*Photograph by Stephen David*

# The Grand Scheme Of Things (A Macrocosmic Perspective)

*Brett Zimmerman*

I'm riding in a supersonic jet
which carries a bomb
towards a city on a continent
in the most important war in
the history of mankind; indeed,
the fate of mankind
is on the line.

But the continent over which
I move at mach speed
is itself moving
eastward from the
mid-Atlantic rift
because of continental drift.

And the earth over which
the continent slides
does itself glide
like an electron around
Sol - the medium-sized
star which is our sun.

Sol itself hurtles through
space in a never-ending journey
towards the blue star
Vega in the constellation
Lyra which is near
Hercules.

These stars in their respective
patterns swirl in their

respective spiral arms
in our whirlpool galaxy
which twirls around like a
great stellar maelstrom.

And this great stellar maelstrom,
this island of stars,
this galaxy,
spins through eternity in its
own cluster of galaxies
which are like burning snowflakes
falling through the vacuum
of the darkness.

And this cluster of galaxies
may be in its own
**supercluster**, which itself
slides and glides,
whirls and swirls,
forever and ever,
amongst all the other superclusters,
**ad infinitum...**

# A Question

*Lee Wallen*

Have you ever felt like a master of your destiny
And that you could change anything that old Zeus
Could throw at you into a piece of easy dancing???
Me neither.

## And When...

*Lee Wallen*

And when the world foresakes you,
I will be here,
For I am faith.

And when mankind betrays you,
I will take your side,
For I am trust.

And when your dreams are shattered,
I will bind them togetherm,
For I am hope.

And when you are troubled,
I will ease the strain,
For I am comfort.

And if you need time to yourself,
I will wait,
For I am patience.

And if you should need any of these things
To help you through your day,
Or to ease the pain of your longest nights,
I will be yours,
For I am love.

*Moon in the Trees, by Noel Napier*

# Morning

*Jacqueline McKay*

Sharp, bright, crisp. Banks enveloped in white fleece; aroma of promise, adventure, rises as the fingers of dawn draw back the covers of dusk. Slowly it begins to show small signs of life; a trail of smoke arises, outlined grey against the newness of the morning, a dog begins to sound somewhere in the distance. From within; a baby wails, coal-black frying pans are heating, spitting protest as they warm; children can be heard scampering, complaining of the cold as they go. All is quiet again; the lyrical chant of prayer can be heard and the house seems to put forth warmth and love. Suddenly, doors are flung wide and children shrieking, stumble into the new day; heavily clad in bright colors against the crisp morning air. They pass quickly, leaving in their wake, footprints, downtrodden fleece: only a memory remains of the newness that was.

# Floating

*Debra VanVeen*

Sit with me now;
Form your lips,
To say good-bye,
Nothing stays the same
No one can take the blame.

Floating.
The hurt in a dream.
Drawing pictures
Of an underwater world

In my mind's eye,
Living pictures,
Underwater sky.

The ache has moved
From heart to mind,
The autumns,
Now flowing under the frozen ice
And I feel like water
At last.

## The Visitor

*Rachael Boles*

Victoria, confined
in massive gilded frame,
looks down, impassive,
on a century not her own.
All straight and starched,
a little plump of late,
yet regal still,
aloof in widow's black.
I try to catch her eye.

"Come down, Victoria, come down!"
She disdains familiarity; I try again.
"I would be honoured, Your Majesty,
if you would consent to take tea with me."
This time she come with rustle of silks
and a faint sigh for her lost love
and her lost world.

But entering my house, her sigh
is of relief. Before her eyes,
Victoriana--carvings and curlicues
and overstuffed chairs, peopled
(surely she can see them)
with comfortable ghosts,
whose hands have touched,
whose hearts have loved,
the things she touched and loved.

I bring in tea. She is
still standing in the middle
of the room, drinking it all in --
burl walnut swirling over table top,
french polished to a righteous glow;
pedestal with barley sugar twist;
fringed lamps and chiffonier.

"And what is this?" She finds
another treasure. "A tea-chest!
My dear, a rosewood tea-chest!
Just like the one I used to have
for locking up my China tea."
She lifts the lid and finds me out --
no tea, just letters tied up with string.

She sits upright on my unforgiving couch
that chides you if you sprawl.
She drinks her tea from a Wedgewood cup.
A smile creases her impassive portrait face.
Here, she is not uncomfortable at all.

## Bonnie Scotland

*Fiona Anderson*

Clootie dumplin's, half a croon
The friendliness of a little toon
Thirty degree weather in the middle of June,
Ye'll find in Bonnie Scotland.

Oor Wullie, The Broons and many more
Have always caused me to laugh and roar
The Dunlop characters I adore,
Ye'll find in Bonnie Scotland.

Woolly sheep, a muckle coo
Old pict houses and tartans true
the things that I've seen noo,
Ye'll find in Bonnie Scotland.

## but for silence

*Catherine O'Toole*

but for silence
do I love
the night.

lines ease
forms melt --
finally

finally

dark.

in the silence
open eyes
are blind.

no one need hear
no one need see --
finally
finally
free.

## Son

*Bill Kischuck*

Trying to break free I fail
Forever running, never moving
I stand dependent in all things
Growing behind bars

What I owe, I cannot repay
Pricked by memory

I refuse to remain inside
Still I stay

Wanting to leap, shamed to safety
I skirt the edge,
Confident of the net that furthers
My entanglement

I shall never rise above
These unyielding clouds
And never shall the sun strike
My careful face.

## The Inevitable

*Minnie Sakhuja*

There's a veil before my eyes
I see things through a misty haze
I look up to the blue skies
And stare with a steady gaze
And see the world
Churn.

Flashes of coloured glass
Kaleidoscope behind my eyes
Slashes of bloody knives
Rip through cerulean skies
And the world
Burns.

Ashes of shattered visions
Block the living sun's fire
A final flickering ember
Glows weakly on a pyre
The world never
Learns.

# Education

*Jim Nolan*

A normal day,
In a normal city.
A classroom,
A university.
He sat
There at his desk.
A normal student.

He listened,
With his normal half attention,
To a man
Who was definitely less than normal.
A fool
Who bantered on about abnormal topics.

They entered,
Silent and stealthful,
The manners of experience.
In his normal daze
His unlooking eyes
Did not detect them.
And their blades were sharp.

He felt no pain
And made no sound
As they removed their treasure
Replaced its covering
And resealed the cavity.
They left with a small grey prize,
And the others
Never noticed that anyone had been there.

Normal days,
In a normal city.
Normal classrooms,
A university.
He sat
There in his normal desk.
A model student.

When his time had come
He listened,
With his normal half attention,
To a man
Who was definitely less than normal.
When his name was called,
He smiled
And was given a rolled parchment.
Graduation...
 With honours.

## A Musing

*Nancy O'Neill*

Amid darkness and dust
and brooms and old coats
Under the stairs
I sit,
Cut off from all others
(Or they from me)
Wishing things were not so,
That I could once again
Rejoin the world of light and air.
But the chaos drives me
back into this secluded lair
this quiet corner
under the stairs.

*Fantasy, by Judith Reid*

*Attack on Hibernation Ship No. 2, by Judith Reid*

# Last Night I Saw A Carpenter

*Michael Anastacio*

Last night I saw a carpenter
behind the cathedral cross
he opened his arms to me
and asked for loneliness
which I granted him freely
I don't enjoy looking at wounds.

# In Words Unknown

*Lee Wallen*

I've searched for a metaphor to describe you--
A word or a phrase that would be my idea of you
That everyone would see and feel as I do.
The flash of your eyes when you laugh, and
The glow of your skin under the moonlight can only
Have true understanding in my mind.
But my efforts to catch the elusive refrains
Of purity, and honesty, and all of your
Innumerable beauties, and to put them
Into mere frailities of symboy are hopeless.
You are the sensations in me.
The rhythmical combat of joy and pain in my heart,
The fear in my mind, and the knowing tremor
In my hand.

# Wretched of the Earth

*Dilip Banerjee*

Those that toil
from dawn to dusk
on meager fields
that they must husk.
That live in squalor;
huts of straw and mud,
living just barely
by their sweat and blood.

They lead their lives -
content, docile,
knowing no other life
or style.
Their bellies raw
to death from birth,
they die unmourned
these wretched of the earth.

*Dilip Banerjee*

I am amazed that
through all the pain
and anger
such feelings could
survived.
And for so long.
It must have
been a hardy strain
of seed
that was sown
to still live.
How sad that only one
seedling took.

*Dust, Dirt, and Mistakes by Thomas Copeland*

## Searching

*Lucille Chenier*

I can hold the world
In the hollow of my mind
Examine its various parts
Follow the flow of water
To the mingling seas.
In the fathomless dark
The universe spins on
Before my unseeing eyes
Determining time and place.
But in the noise of day
Or the quiet of night
I cannot find the essence
Of you.

## Here in the Valley

*Michael Goodfellow*

Here in the valley
That time left behind
That the glaciers carved
That the river flows through
Where the trees grow strong
And the winters are long
Lies the knowledge of the world
Laid out for all to see
And for those who wish, to grasp.

## even if

*Catherine O'Toole*

the sun is not today bright enough
glancing off the snow to
sting
my eyes.
but the wind is cold enough
glancing off the sky to
sting me
outside,
if i could be

outside,
could feel the sting;
could feel the cold.
Even if it made me cry - at least
i'd know it was real.

my love is not today wise enough
glancing off you to
shade
my heart.
but your actions are cold enough
glancing off the shadows to
darken me
inside,
if i could be

inside,
i could light the shadows;
i could face the cold.
even if it made me cry - at least
i'd know you are real.

# On Being Handed Poems on a Plate

*Rachael Boles*

So you labour all day
over a poem
mixing it
shaping it
tossing it
stretching it
rhyming it
timing it
writing and rewriting

Sauce
seasoned with subtlety
laid over with
slices of wisdom
and grated wit
Should be good
you think
Homemade and all

Then Inspiration
sends a poem
ready made
to your door
and taste tests show
that sometimes
the take-out poems
are better than the ones
you cook up yourself
at home

# index

All inquiries should be addressed to:

The Editor
c/o Scarborough College Students' Council,
Scarborough College
University of Toronto
1265 Military Trail
West Hill, Ontario,
M1C 1A4.

The material in this anthology has been selected from over three hundred and fifty submissions that were contributed by eighty-five writers and artists.